Beekeeping for Beginners
Linda Standridge

Beekeeping for Beginners

Linda Standridge

Published by Linda Standridge, 2023.

While every precaution has been taken in the preparation of this book, the publisher assumes no responsibility for errors or omissions, or for damages resulting from the use of the information contained herein.

BEEKEEPING FOR BEGINNERS

First edition. October 22, 2023.

ISBN: 979-8223564553

Written by Linda Standridge.

Table of Contents

Chapter One

The History of Beekeeping: an Introduction

In order to truly understand the fascinating world of beekeeping, we must delve into its rich history. The practice of beekeeping, or the management of bees for the production of honey and other bee products, has been an integral part of human civilization for centuries. From ancient civilizations to modern times, beekeeping has played a significant role in our lives. Beekeeping can be traced back to ancient Egypt, where evidence of early beekeeping practices can be found in hieroglyphs and ancient artworks.

The Egyptians revered bees and their honey, not only for it's delicious taste but also for its believed medicinal and spiritual properties. They even considered bees to be sacred creatures, often associating them with their deities.

Moving forward in time, we come across the ancient Greeks and Romans, who were also avid beekeepers. The Greeks, in particular, recognized the importance of bees for their role in pollination, which directly impacted agriculture and the growth of their civilization. They were the first to document systematic beekeeping techniques and even developed beehives designed to facilitate honey production.

During the Middle Ages, beekeeping continued to evolve, with advancements in hive design and beekeeping practices. Monasteries played a crucial role in cultivating and spreading beekeeping knowledge. Monks recognized the economic and nutritional value of honey and beeswax, and thus beekeeping became an essential part of monastic life.

The practice of beekeeping expanded further during the Renaissance, as the study of natural sciences flourished. Scholars like Jan Dzierzon and Lorenzo Langstroth made significant contributions to beekeeping by developing innovative hive designs and techniques that revolutionized the industry. These advancements allowed beekeepers to better manage their colonies, leading to increased honey production and better overall hive health.

In more recent times, beekeeping has become a popular hobby and a vital industry worldwide. The understanding of bee behavior, the role of pollinators, and the importance of honey production have continued to grow. Today, beekeepers employ various methods to ensure the well-being of their colonies and to maintain the delicate balance between human needs and ecological sustainability. As we embark on this journey through the history of beekeeping, we will explore the ancient practices, the technological advancements, and the cultural significance of this remarkable craft. We will uncover the stories of the beekeepers who dedicated their lives to understanding and nurturing these incredible creatures. Through their efforts, we have come to appreciate the vital role that bees play in our ecosystem and the countless benefits they provide. So, let us delve into the captivating history of beekeeping, an introduction to the world that revolves around these small but mighty creatures. Together, we will uncover the secrets of their ancient origins and witness the ongoing evolution of beekeeping practices. Through this exploration, we will gain a deeper appreciation for the intricate relationship between humans and bees, prompting us to protect and cherish these remarkable insects for generations to come

Chapter Two

Bees and Their Anatomy

We will delve into the intricate and remarkable anatomy of these incredible creatures. It is through an understanding of their anatomy that we can truly appreciate the wonders of beekeeping and the vital role bees play in our ecosystem. Let us begin our exploration by focusing on the body structure of the bee. Bees, like other insects, possess an exoskeleton, which serves as their external protective covering. This exoskeleton is composed of a tough substance called chitin, providing both support and defense for the bee's delicate inner organs.

Moving on to the head, we encounter some remarkable features. Bees have a pair of large compound eyes, allowing them to see the world in a mosaic-like fashion. These compound eyes are made up of thousands of individual lenses, enabling bees to detect motion and perceive ultraviolet light, which is invisible to human eyes. Additionally, bees possess three simple eyes, or ocelli, located on the top of their heads. These ocelli are responsible for sensing changes in light intensity and direction, aiding bees in navigation.

The bee's mouthparts are well adapted for their feeding habits. They possess a long, tube-like structure called a proboscis, which allows them to suck nectar from flowers and collect pollen. This proboscis is formed by a fusion of the bee's maxillae and labium, creating a specialized tool for their unique dietary needs.

Next, we move down the thorax, the middle section of the bee's body. Here, we find a fascinating arrangement of muscles that control the bee's wings. Bees have four wings in total, with the forewings being larger than the hindwings. Through precise muscle contractions, bees are able to achieve the rapid wing movements necessary for flight, hovering, and intricate aerial maneuvers.

Finally, we come to the abdomen, which houses several vital organs. The bee's digestive system, including the crop and the stomach, is located here, allowing for efficient processing of nectar and pollen. The abdomen also contains the wax glands, responsible for producing the wax that bees use to construct their intricate honeycombs.

Additionally, bees have specialized glands, such as the venom gland and the scent gland, which play crucial in defense and communication. Understanding the anatomy of bees is essential for any beekeeper. By gaining knowledge of their unique structures and functions, we can better appreciate the incredible adaptations that have allowed bees to thrive for millions of years. It is through this understanding that we can continue to support and protect these vital pollinators, ensuring the health and sustainability of our ecosystems. Remember, in the world of bees, their anatomy is truly a marvel to behold.

Chapter Three

The Life Cycle of the Bee

The life cycle of the bee is a fascinating journey that encompasses various stages. Understanding this cycle is crucial for any beekeeper or enthusiast, as it provides valuable insight into the intricate world of these remarkable creatures. It all begins with the queen bee, the central figure in the hive. She is responsible for laying eggs, which will eventually develop into new bees. The queen's primary task is to ensure the survival and growth of the colony by producing offspring. The queen bee lays her eggs in specially constructed cells within the hive.

These cells are meticulously crafted by worker bees, who use beeswax to build the hexagonal structures known as honeycombs. Each cell is designed to hold a single egg, and the queen strategically places them throughout the hive. Once the eggs are laid, they undergo a process known as metamorphosis. This transformation is truly remarkable, as it involves three distinct stages: the egg, the larva, and the pupa. Each stage serves a unique purpose in the development of the bee.

First, the egg hatches, giving birth to a tiny larva. The larva is entirely dependent on the worker bees for its nourishment. They diligently feed the larva with a substance called royal jelly, which is a nutritious secretion produced by worker bees. This royal jelly is essential for the larva's growth and development. As the larva continues to feed on royal jelly, it grows rapidly, shedding its skin several times. During this phase, the larva is often referred to as a "nurse bee" due to its reliance on the worker bees for sustenance. The worker bees are responsible for maintaining the ideal temperature and humidity in the hive to ensure the larva's well-being.

After a few days, the larva enters the pupa stage, where it undergoes profound changes. During this stage, the larva is encased in a protective cocoon, which shields it from external elements. Inside the cocoon, the larva goes through a remarkable transformation, gradually developing into an adult bee.

Finally, after a few weeks, the adult bee emerges from the cocoon, ready to join the hive. This stage marks the completion of the life cycle and the beginning of the bee's role in the colony. The newly emerged bee's tasks may vary depending on its gender and role in the hive. Understanding the life cycle of a bee is crucial for beekeepers, as it allows them to anticipate and manage the needs of the hive. By comprehending the different stages, beekeepers can ensure the well-being and productivity of their colonies.

In conclusion, the life cycle of the bee is a complex and fascinating process that encompasses the journey from egg to adult. This cycle is vital for the survival and growth of the hive, and understanding it is key to successful beekeeping. By delving into the intricacies of this cycle, we gain a deeper appreciation for the remarkable world of bees.

Chapter Four

The History of Beekeeping

The history of beekeeping is a fascinating tale that spans thousands of years. From ancient civilizations to modern times, the art of beekeeping has evolved and shaped our understanding of these vital insects. In the earliest days of beekeeping, our ancestors discovered the value of honey as a natural sweetener and source of energy. Primitive methods involved gathering honey from wild bees, often risking encounters with aggressive bees. True beekeeping began when humans started providing artificial constructs where the bees could build a honeycomb for the queen to lay her eggs and the workers could process honey.

Beekeeping is an ancient craft in both the Old and New Worlds. The oldest known Old World bee hives are from Tel Rehov, in the current state of Isreal, about 900 B.C.E; the oldest known in the Americas is dated from between 300 B.C.E.-200/250 C.E. which is known as the the Late Preclassic or Protoclassic period Maya site of Nakum, in the Yucatan peninsula of Mexico. The Maya cultivated a stingless bee known today as Melipona beecheii.

Over time, humans began to realize the importance of managing these hives and nurturing the bees within. Beekeeping in India is mentioned in ancient scriptures known as the Vedas and in Buddhist scriptures. Rock paintings from the Mesolithic era found in Madhya Pradesh depict honey collection. One of the earliest known records of intentional beekeeping dates back to ancient Egypt. Hieroglyphics found in tombs depict scenes of beekeepers tending to their hives and collecting honey. The Egyptians revered bees and associated them with their gods, recognizing the incredible work these insects put into creating honey. By 2450 B.C.E. the Egyptians had developed a sophisticated apiculture. By 450 B.C.E. beekeeping by the use of horizontal hives had spread out across the Mediterranean. As civilizations expanded and trade routes flourished, beekeeping techniques spread across the world.

In China, archaeological evidence for human consumption of honey dates to the seventh millennium BCE. However, the earliest reference to beekeeping is a third-century CE biography of a mid-second-century reclusive scholar who kept bees. It discusses the methods used by Chinese beekeepers to try to increase production, harvest honey, control swarming, and maintain colonies over the winter, as well as their understanding of bee behavior in relation to the governance of the colony and the production of honey.

The Greeks and Romans embraced beekeeping as well, recognizing the value of honey not only as a food source but for its medicinal properties. Greek philosopher Aristotle even wrote about the behavior and habits of bees, laying the groundwork for future beekeeping knowledge.

During the Middle Ages, beekeeping became a common practice in monasteries and religious institutions. Monks saw bees' industrious nature as a reflection of their own dedication and hard work. They developed innovative hive designs, such as the skep, a domed straw basket that allowed for easier honey extraction. In those days, Everyone believed a "king bee", not a "queen bee", ruled the swarm. During the latter half of the 12th century, beekeeping came to Japan.

The Renaissance period brought advancements in beekeeping technology. European farmers started using movable frame hives, which allowed for better hive management and honey extraction without disturbing the bees. These developments laid the foundation for modern beekeeping methods still in use today.

In the 19th century, the Industrial Revolution brought about further innovations in beekeeping. The invention of the honey extractor revolutionized honey collection, making it faster and more efficient. Beekeepers began to understand the intricate dynamics of the beehive, leading to improved colony management practices.

The 20th century witnessed significant advancements in beekeeping techniques and research. Beekeepers started to focus on breeding healthier and more productive bee colonies. The discovery of the role of the queen bee in colony development led to selective breeding programs, aimed at producing bees with desirable traits.

However, the history of beekeeping is not without its challenges. In recent decades, the global bee population has faced numerous threats, including habitat loss, pesticide use, and climate change. These factors have led to declines in bee populations worldwide, raising concerns about the future of beekeeping and the environmental impact of losing these vital pollinators.

Today, beekeeping continues to be an essential practice for both hobbyists and commercial beekeepers. The importance of bees in pollination cannot be overstated, as they contribute to the reproduction of countless species, including important food crops. Beekeepers play a crucial role in maintaining healthy bee populations and safeguarding the future of our ecosystems. The history of beekeeping is a testament to humanity's fascination with and reliance on these remarkable creatures. From ancient civilizations to our modern world, the art of beekeeping has evolved, adapted, and continues to thrive. As we navigate the challenges of the present and look toward the future, it is our responsibility to preserve the legacy of beekeeping, ensuring the survival of these extraordinary insects for generations to come.

Chapter Five

Beekeeping Practices and Equipment

Beekeeping has a rich history that dates back centuries, with practices and equipment evolving over time to meet the needs of the beekeepers. We will delve into the fascinating journey of beekeeping practices and the essential equipment used in this ancient craft. The art of beekeeping has been practiced since ancient times, with evidence of its existence found in various cultures around the world. In Ancient Egypt, for example, beekeeping was depicted in hieroglyphics, showing the importance of bees and honey in their society. Ancient Greeks and Romans also recognized the significance of bees, using them for honey production and pollination of crops.

Throughout history, beekeepers have developed various practices to maintain healthy and thriving bee colonies. One of the earliest methods involved keeping bees in hollowed-out logs or baskets, allowing them to build their honeycomb freely. This practice, known as skep beekeeping, was prevalent in Europe for centuries. As the understanding of bees and their behavior grew, beekeepers began to develop more sophisticated practices.

The invention of movable frames in the 19th century revolutionized beekeeping. These frames allowed beekeepers to inspect the hives without disturbing the bees excessively. Beekeepers could not manipulate the frames to encourage specific behavior, such as honey production or queen rearing. Modern beekeeping practices focus on maintaining the health and well-being of the bees while maximizing honey production.

Beekeepers employ techniques like integrated pest management (IPM) to control pests and diseases without relying heavily on chemical treatments. They also strive to provide a diverse forage environment for their bees, ensuring a steady supply of nectar and pollen throughout the year.

To facilitate these practices, a range of beekeeping equipment has been developed. The most recognizable piece of equipment is the beehive itself. Modern beehives are designed to provide a suitable habitat for the bees, allowing easy access to honey and pollen stores while protecting them from the elements. Within the hive, beekeepers use frames and foundation sheets to guide the bees comb-building activities. Frames provide support for the comb, making it easier for beekeepers to inspect and manipulate the hive. Foundation sheets, made of beeswax or plastic, provide a blueprint for the bees to build their comb, ensuring straight and uniform cells. Other essential equipment includes the smoker, which produces smoke that calms the bees during hive inspections, and the bee suit, which protects the beekeepers from stings. Beekeepers utilize tools like hive tools, bee brushes, and honey extractors to handle and collect honey from the hive.

In recent years, advancements in technology have further enhanced beekeeping practices. Beekeepers can now monitor hive conditions remotely using sensors and data analysis, allowing for quick intervention in case of any issues. This technology enables beekeepers to provide the best possible care for their bees and optimize honey production.

In conclusion, beekeeping practices and equipment have come a long way throughout history. From the humble skep to modern beehives and advanced monitoring systems, beekeepers have continually adapted their methods to ensure the well-being of their bees and the production of high-quality honey. By understanding the historical contact and embracing innovative techniques, beekeepers can continue this ancient tradition while promoting the sustainability of the beekeeping industry.

The Role of Bees in Pollination

Throughout history, the significance of bees in pollination has been paramount. Bees, these incredible creatures, have played a vital role in the propagation of plants and the sustenance of our ecosystems. We will delve into the fascinating history and explore how bees have become indispensable agents of pollination. The practice of beekeeping, known as apiculture, dates back thousands of years.

Ancient civilizations, such as the Egyptians, the Greeks, the Chinese, and the Maya, recognized the value of bees and their role in pollination. They observed the natural connection between bees and the growth of plants, realizing the process of pollination was crucial for the production of fruits, vegetables, and grains. In ancient Egypt, for example, beekeeping was not only a means to obtain honey but also a way to ensure successful pollination of vital crops such as figs, dates, and pomegranates. The Egyptians even depicted bees and beekeeping in their hieroglyphs, emphasizing the importance they placed on these remarkable insects.

The Greeks, too, held bees in high regard. They believed that bees were sacred creatures, connected to the gods. Greek mythology tells the tale of Aristaeus, the god of beekeeping, who was responsible for teaching humans the art of apiculture. This mythological association further emphasized the crucial role of bees in the natural world.

As time passed, beekeeping evolved and spread across continents. In Medieval Europe, monastic orders took up beekeeping as a means of sustaining themselves and their communities. Monasteries became the centers of knowledge and innovation, where monks experimented with different hive designs and techniques to improve honey production. Little did they know that their efforts would also contribute to the health and diversity of nearby plant life through pollination.

With the arrival of the Age of Enlightenment, scientific curiosity took hold, and the study of bees became more systematic. In the 18th century, the Swedish scientist Carl Linnaeus classified and named thousands of species, including bees. His work laid the groundwork for understanding the intricate relationships between different bee species and the plants they pollinate. In more recent times, the role of bees in pollination has become even more critical. As human populations have grown, so too has the demand for food.

Bees, as efficient pollinators, have become essential for maintaining agricultural productivity. They are responsible for pollinating around 75% of the world's crops, including fruits, vegetables, and nuts. However, in the face of environmental challenges such as habitat loss, pesticide use, and climate change, bees and their ability to pollinate are under threat.

The decline of bee populations, particularly honeybees, has raised concern about the future of pollination and food security. Understanding the role of bees in pollination is not only a fascinating journey through history but also a call to action. It is crucial that we protect and conserve bee populations, ensuring their habitats remain intact and reducing the use of harmful chemicals. By doing so, we can safeguard their vital role in pollination and maintain the delicate balance of our ecosystems.

We will explore the intricate dance between bees and plants, the fascinating mechanisms of pollination, and the extraordinary symbiotic relationships that exist in the natural world. By delving deeper into the role of bees in pollination, we can gain a greater appreciation for these remarkable creatures and work towards a future where their invaluable contribution is secured for generations to come.

Chapter Seven

Bee Products and Their Uses

Bee products and their uses have been an integral part of human history for thousands of years. The practice if beekeeping, or apiculture, dates back to ancient times and has played a significant role in civilizations across the globe. In order to understand the significance of bee products and their uses, it is essential to delve into the rich history of beekeeping. The origins of this practice can be traced back to ancient Egypt, where honey was prized for its medicinal properties and was used as an offering to the gods. The Egyptians even depicted beekeeping scenes on their temple walls, highlighting the importance they placed on this craft.

Moving forward in time, we find that beekeeping was also prevalent in ancient Greece and Rome. These civilizations recognized the value of bee products, particularly honey and beeswax. Honey was used not only as a sweetener but also as a preservative, and it was highly sought after for its taste and health benefits. Beeswax, on the other hand, was used for a variety of purposes, such as making candles, cosmetics, and even a waterproofing agent.

Beekeeping techniques and practices continued to evolve throughout the Middle Ages and into the Renaissance period. Monasteries became centers of beekeeping knowledge, and monks played a crucial role in advancing the craft. They experimented with hive design, improved honey extraction methods, and contributed to the understanding of bee behavior.

As the Age of Exploration dawned, beekeeping spread to new territories. European settlers brought their knowledge of beekeeping to the Americas, where indigenous people also had a long history of utilizing bee products. The Native Americans, for instance, used honey as a natural sweetener and incorporated it into various traditional recipes.

In more recent times, beekeeping has become increasingly important for its role in pollination. Bees play a vital role in the pollination of a wide range of crops, ensuring food production and agricultural sustainability. With the decline of the wild bee population, the practice of beekeeping has taken on a greater significance in preserving biodiversity and supporting ecosystem health.

Today, the uses of bee products extend beyond honey and beeswax. Bee pollen, propolis, and royal jelly have gained recognition for their potential health benefits. Bee pollen is a nutrient-rich superfood, propolis is known for its antimicrobial properties, and royal jelly is believed to have anti-aging effects. These products have found their way into dietary supplements, skincare products, and alternative medicine practices.

In conclusion, bee products and their uses have a long and fascinating history that spans across cultures and centuries. From ancient civilizations to modern times, beekeeping has provided us with valuable products that have been utilized for their culinary, medicinal, and industrial properties. As we navigate the challenges of the present, it is crucial to continue supporting the practice of beekeeping, recognizing the importance of bees, and harnessing the potential of bee products for the benefit of both humans and the environment.

Chapter Eight

Bee Disease and Pests

Bee disease and pests have long plagued the practice of beekeeping, leaving beekeepers to grapple with the challenges of the present. Understanding the history of beekeeping allows us to delve into the origins of these issues and shed light on the evolution of our efforts to combat them. Throughout the ages, beekeepers have encountered various diseases and pests that have threatened the health and productivity of their colonies. The earliest recorded instances of these afflictions date back thousands of years, highlighting the enduring nature of these challenges. In ancient Egypt, for instance, beekeepers faced the wrath of foulbrood disease. This devastating bacterial infection could decimate entire hives, leaving beekeepers with no choice but to burn affected colonies to prevent the spread of the disease. Although they lack the scientific understanding we possess today, these early beekeepers recognized the importance of quarantine and eradication measures.

Moving forward in time, we come across the emergence of another notorious bee disease: American foulbrood. This highly contagious bacterial infection was first identified in the United States in the late 19th century. With its ability to spread rapidly and persist in hives, American foulbrood posed a significant threat to beekeeping industries worldwide. The discoveries of antibiotics and the development of effective treatment methods marked a turning point in our battle against this disease.

As we delve deeper into the history of beekeeping, we encounter an array of pests that have plagued bee colonies throughout the centuries. The varroa mite, for example, entered the scene in the mid-20th century, wreaking havoc on bee populations. This parasitic mite feeds on adult bees and their brood, weakening colonies and leaving them susceptible to further disease and colony collapse.

The small hive beetle, another formidable adversary, made its presence known in the late 20th century. Originating in Africa, this invasive pest quickly spread to the other continents, causing damage to honeycombs, contaminating honey, and disrupting the delicate balance within the hive.

Over time, beekeepers have developed various strategies to mitigate the impact of these pests and diseases. From selective breeding for disease-resistance traits to the use of chemical treatments and integrated pest management approaches, beekeepers have tirelessly sought solutions to protect their beloved colonies.

Today, advancements in research and technology continue to offer hope for the future of beekeeping. Scientists are exploring alternative treatments, such as biological controls and genetic modifications, to combat diseases and pests effectively. Additionally, an improved understanding of bee biology and behavior allows beekeepers to implement proactive measures to prevent the spread of ailments.

As we reflect on the history of beekeeping, it becomes evident that bee diseases and pests have been persistent adversaries. Nonetheless, the ingenuity and dedication of beekeepers have led to significant strides in disease management and pest control. By learning from the past and embracing new knowledge, we can forge a path toward sustainable beekeeping practices that ensure the health and vitality of these vital pollinators.

Chapter Nine

Honey Bee Nutrition

Honey Bee nutrition plays a vital role in the history of beekeeping. Throughout the ages, beekeepers have recognized the importance of nourishment to their honey bee colonies. This chapter delves deep into the fascinating journey of understanding honey bee nutrition and its impact on the development of beekeeping practices. From ancient civilizations to modern times, beekeepers have observed and experimented with various sources of nutrition for their honey bees.

In the early days, beekeepers relied on the natural flora surrounding their hives, allowing their bees to forage on a diverse range of nectar and pollen sources. They understood that a balanced diet was crucial for the health and productivity of their colonies. As beekeeping evolved, so did our understanding of honey bee nutrition.

In the late 19th century, advancements in scientific research and the discovery of the role of carbohydrates, proteins, lipids, vitamins, and minerals in honey bee health revolutionized our approach to beekeeping. Beekeepers began to supplement their bee's diet with sugar syrup, pollen substitutes, and protein supplements to ensure their colonies had access to a well-rounded nutritional profile. The importance of honey bee nutrition became even more evident during times of food scarcity, such as winter months or when colonies were moved to areas with limited forage resources. Beekeepers started to develop feeding strategies to provide their bees with essential nutrients during these challenging periods, ensuring the survival and strength of their colonies. However, it is crucial to understand that honey bee nutrition is not just about providing the bare minimum to sustain the colonies. Proper nutrition goes beyond survival; it contributes to the overall health, immune system strength, and vitality of the honey bee population. A well-nourished colony is more resistant to diseases, pests, and environmental stresses.

In recent years, extensive research has been conducted to unravel the intricate relationship between honey bee nutrition and their overall well-being. Scientists and beekeepers alike have come to realize the importance of a diverse and abundant floral landscape for honey bee nutrition. Monoculture agricultural practices, pesticide use, and habitat loss have had a significant impact on the availability and quality of honey bee forage. As beekeepers, it is our responsibility to ensure our honey bees have access to a wide variety of nutritious floral resources throughout the year. Planting diverse flowering plants, avoiding pesticide use, and providing supplemental nutrition when necessary are crucial steps in promoting honey bee health and longevity.

In conclusion, the history of beekeeping is closely intertwined with the understanding and implementation of honey bee nutrition. From ancient civilizations to modern scientific advancements, beekeepers have recognized the importance of providing their colonies with a balanced and diverse diet. As we continue to learn and adapt our practices, let us remember that honey bee nutrition is not just a chapter in the history of beekeeping; it is an ongoing journey toward the well-being of these incredible pollinators.

Chapter 10

The Future of Beekeeping

In exploring the rich history of beekeeping, it becomes evident that this age-old practice has evolved and adapted throughout the ages. From ancient civilizations to modern times, humans have recognized the invaluable role bees play in our ecosystems and our lives. However, as we delve deeper into the annals of beekeeping, we also come face to face with the challenges and uncertainties that lie ahead. The future of beekeeping holds both promise and concern.

As we stand on the cusp of a new era, it is essential to examine the factors that will shape the landscape of beekeeping in the years to come. One of the key elements influencing the future of beekeeping is the ever-growing awareness of the crucial role bees play in pollination and biodiversity. With the decline of bee populations in recent years, the urgency to protect these vital creatures has become paramount.

Governments, environmental organizations, and individuals have rallied together to implement measures aimed at preserving and restoring bee habitats. This collective effort is a testament to the recognition of the significance of beekeeping not only for honey production but also for the overall health of our planet.

The future of beekeeping is intricately intertwined with technological advancements. The advent of innovative tools and techniques has revolutionized various aspects of beekeeping practices. From hive monitoring systems that provide real-time data on hive conditions to the development of more efficient extraction methods, technology has the potiential to enhance productivity and optimize beekeeping operations. Moreover, the utilization of artificial intelligence and machine learning algorithms can aid in predicting hive health and identifying potential threats. By harnessing these technologies, beekeepers can detect diseases or environmental factors that may be detrimental to bee colonies in a timely manner, enabling them to take appropriate measures to protect and support their bees.

Another vital aspect to consider when contemplating the future of beekeeping is the changing landscape of agriculture. As farming practices evolve to meet the demands of a growing global population, so too must beekeeping adapt. With the expansion of monoculture and the increased use of pesticides, bees face new challenges in finding diverse and nutritious forage. To overcome these obstacles, beekeepers and farmers must collaborate to create bee-friendly landscapes. The integration of pollinator-friendly plants in agriculture fields and the implementation of sustainable farming practices can provide bees with a more diverse and abundant food supply. This partnership between beekeepers and farmers is essential for the long-term survival and success of both bees and agriculture.

Education and awareness will also play a crucial role in shaping the future of beekeeping. By educating the public about the importance of bees and the threats they face, we can foster a sense of responsibility and encourage individuals to take action. Schools, community organizations, and beekeeping associations can all contribute to spreading knowledge and empowering citizens to become advocates for the bees.

In conclusion, the future of beekeeping is a complex tapestry woven with challenges and opportunities. By prioritizing the protection of the bees, embracing technological advancements, fostering collaboration between beekeepers and farmers, and promoting education, we can create a future where beekeeping thrives. Together, we can ensure that the invaluable role of bees in our ecosystems continues to be cherished and protected for generations to come.

Chapter Eleven

Introduction to the Basics

Beekeeping is an ancient practice that involves the care and management of bees for the purpose of harvesting honey and other bee products. It is a fascinating and rewarding hobby that allows us to connect with nature while also contributing to the health of our ecosystems. We will explore the fundamentals of beekeeping, providing you with a solid foundation to embark on the bee-filled journey.

First and foremost, it is essential to understand the importance of bees in our world. Bees are not just honey producers; they are crucial pollinators that play a vital role in the reproduction of flowering plants. Without bees, many of the fruits, vegetables, and nuts that we rely on for sustenance would cease to exist. By becoming a beekeeper, you are actively participating in the preservation of these vital pollinators and the biodiversity of our planet.

Before delving into the practical aspects of beekeeping, it is important to familiarize yourself with the different types of bees you may encounter. The most common species kept by beekeepers is the Western honey bee (Apis Mellifera). These bees are known for their gentle nature and high honey production.

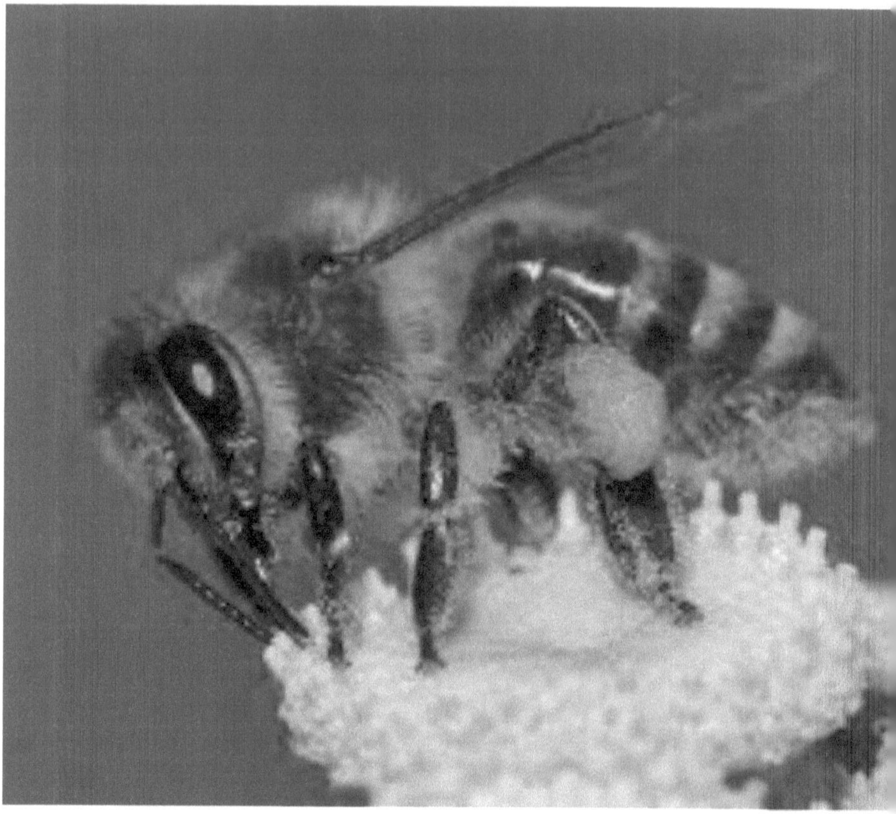

Apis Mellifera (Western honey bee)

HOWEVER, THERE ARE also other species, such as the African honey bee, which have different characteristics and may require specific management techniques. Africanized honey bees are dangerous because they attack intruders in numbers much greater than European honey bees. Victims receive ten times more stings than you would typically get from the European strain. They react much quicker than European honey bees and they will chase a person a quarter of a mile. Other concerns are the damages done to the honey industry, annual damage at a cost of $140 million dollars. And the effects on the pollination of orchards and food crops, an annual loss of crops cost 10 billion dollars. Africanized honey bees have been known to attack and kill hives of Western honey bees.

Africanized honey bees are slightly smaller than Western honey bees. They are golden yellow with darker bands of brown.

Once you have identified the type of bee you wish you to keep, the next step is to acquire the necessary equipment. Essential beekeeping tools include a bee suit, gloves, a smoker, hive tools, and bee brushes. These items will help protect you from stings and allow you to handle the bees safely and efficiently. Additionally, you will need beehives, frames, and foundation sheets to provide a suitable habitat for your bees.

Beehive in a wildflower meadow

Frames for beehive

Beekeeper's suit

Beekeeping gloves

Foundation sheets

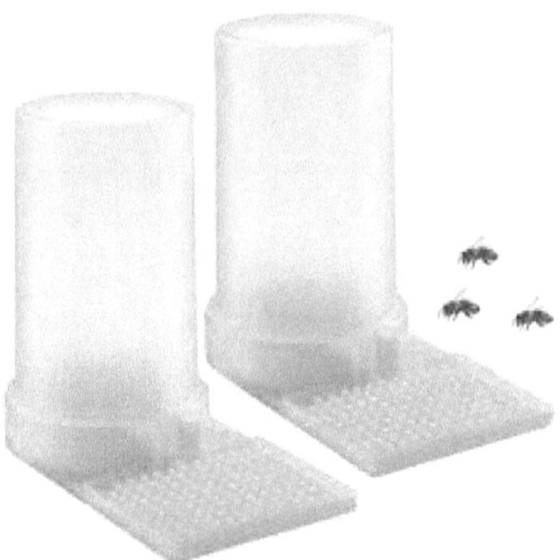

Bee feeder

Honey extractor

HIve tool

bee brush

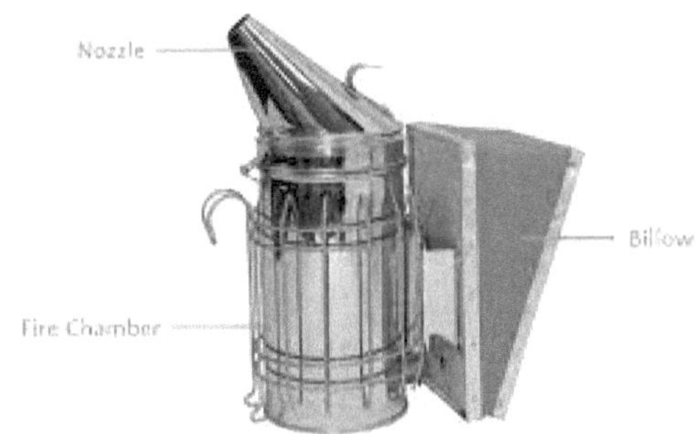

bee smoker

ONE CRUCIAL ASPECT of beekeeping is selecting the right location for your hives. Bee hives thrive in areas with abundant nectar sources, so it is important to choose a site that offers a diverse range of flowering plants, Additionally, the location should be easily accessible for you to monitor and maintain your hives. Consider the availability of water sources and the proximity to potential disturbances, such as livestock or busy roads.

Now that you have the necessary equipment and have chosen a suitable location, it is time to introduce bees into your hives. The most common method is to purchase a package of bees, which includes a queen and a group of worker bees. Another option is to catch a swarm, which is a natural process where bees split from an existing colony and form a new one. Whichever method you choose, it is crucial to handle the bees with care and follow proper procedures to ensure a successful introduction.

Once your bees are settled in their new hives, it is important to regularly inspect and maintain them. This involves checking for signs of diseases or pests, monitoring honey production, and providing supplemental food if needed. Regular hive inspections will allow you to address any issues promptly, ensuring the health and productivity of your bee colonies. Beekeeping is a lifelong learning journey, and this chapter merely scratches the surface of its vast knowledge.

As you delve deeper into the world of beekeeping, you will encounter various techniques, challenges, and rewards. Remember to approach beekeeping with patience, respect, and a willingness to learn from both successes and failures. In the following chapters, we will explore the intricacies of hive management, honey harvesting, and bee health. By the end of this book, you will have a comprehensive understanding of beekeeping basics, empowering you to embark on your own beekeeping adventure. So, let us dive in and explore the fascinating world of beekeeping together.

Chapter Twelve

Honey Bee Anatomy, Taxonomy, and Behavior

In order to understand the art of beekeeping, it is essential to delve into the intricate world of honey bee anatomy, taxonomy, and behavior. By exploring the various aspects of their physical structure and classification, we can gain invaluable insights into their behavior and needs of these remarkable creatures. Bees are not only remarkable creatures but they also play a crucial role in our ecosystem as pollinators. Understanding their biology and behavior is essential for any aspiring beekeeper or enthusiast.

Let us begin by examining the anatomy of honey bees. Bees are insects with three main body segments: the head, thorax, and abdomen. The body is covered in dense hairs, which aid in the collecting of pollen and distributing it during pollination. At first glance, the honey bee may appear simple, but upon close inspection, we find a complex and fascinating organism.

Starting with the head, we find the sensory organs that enable honey bees to navigate and communicate within their environment. Their large compound eyes provide them with a wide field of vision, allowing them to detect movements and patterns in their surroundings. Their compound eyes provide excellent vision and their antennae play a vital role in sensing their environment. In addition, honey bees possess three sets of eyes, or ocelli, situated on the top of their heads. The Ocelli play a crucial role in helping bees orient themselves to the sun and maintain stability during flight.

Moving down to the mouthparts. Honey bees are equipped with various tools for gathering nectar and pollen. Their long proboscis, allows them to access the nectar hidden deep within flowers. Alongside their tongues, they have specialized mouthparts that aid in the collection and manipulation of pollen, an essential source of protein for the hive.

Next, we come to the thorax, which is the central region of the honey bee's body. Attached to the thorax are three pairs of legs, each uniquely adapted to fulfill different tasks. The hind legs, in particular, possess specialized structures known as pollen baskets or corbiculae. These baskets are concave areas surrounded by stiff hairs, enabling bees to collect and transport pollen back to the hive. Situated on the thorax are the delicate and translucent wings that enable honey bees to fly. These wings are composed of a network of veins, providing strength and flexibility during flight. The ability to fly is crucial for honey bees, as it allows them to forage for food, gather resources, and explore their surroundings. They have two pairs of wings, allowing them to fly with exceptional agility and precision.

Lastly, we reach the abdomen, which houses the vital organs and reproductive system of the honey bee. Within the abdomen, the bees' digestive and respiratory systems play a crucial role in maintaining their overall health and well-being. Additionally, the sting apparatus is composed of a modified ovipositor, which is found at the end of the abdomen. Although primarily used in defense, the sting can also be employed during aggressive encounters with rival bees.

Beyond anatomy and behavior, understanding the taxonomy of the honey bee is equally important. The honey bee, scientifically known as Apis Mallifera, belongs to the insect order Hymenoptera, which includes ants, wasps, and sawflies. With this order, bees are classified under the family of Apidae. Within the genus Apis, there are 20,000 recognized subspecies, each with its own unique characteristics and adaptations.

In terms of reproduction, bees are eusocial insects, meaning that they live in highly organized in distinct castes. Each colony consists of a queen, drones, and worker bees. The queen is responsible for laying eggs and maintaining the unity of the colony. Drones are male bees whose sole purpose is to mate with the queen. Worker bees are sterile females who perform various tasks within the colony, such as foraging, nursing the brood, and building and maintaining the hive.

Bee behavior: Now, let us explore the fascinating behavior of bees. Bees are known for their complex social structure and remarkable communication abilities. They communicate through a combination of chemical signals, movements, and sounds. Their ability to communicate effectively is crucial for tasks such as foraging, defending the hive, and coordinating collective decision-making.

One of the most well-known behaviors of bees is their role as pollinators. As bees forage for nectar and pollen, they inadvertently transfer pollen grains from one flower to another, enabling plants to reproduce. This mutualistic relationship between bees and flowering plants is essential for the survival of both parties. Bees also exhibit remarkable navigation skills. They use various cues, including the position of the sun, landmarks, and even the Earth's magnetic field, to navigate accurately and return to their hive. This ability is particularly impressive considering their small size and the vast distances they often cover during foraging.

Another fascinating behavior of bees is their collective decision-making behavior. When a colony needs to find a new site, workers perform a "waggle-dance" inside the hive to communicate information about the location and the quality of potential sites. Through this dance, bees convey directional and distance information to guide their fellow workers to the desired location.

By studying and appreciating the intricate details of honey bee anatomy, behavior, and taxonomy, we can develop a deeper understanding of these incredible creatures. This knowledge serves as a foundation for effective beekeeping practices, ensuring the health and well-being of honey bees in our care. By gaining insight into their intricate social structure, remarkable communication skills, and vital role as pollinators, we can develop a deeper appreciation for these remarkable creatures. As we continue our journey into the world of beekeeping, let us remember the importance of bee biology and behavior in fostering a harmonious relationship between humans and bees. So, let's dive into the captivating world of honey bee anatomy and taxonomy, and unlock the secrets that will enrich our journey as beekeepers.

Chapter Thirteen

Hive Design and Structure

In the fascinating world of beekeeping, understanding the intricacies of hive design and structure is crucial for successful beekeeping. The hive is the home and sanctuary for our buzzing friends, providing them with shelter, security, and a conducive environment for their natural activities. As beekeepers, it is our responsibility to ensure that their hives are well-designed and carefully constructed.

One of the most commonly used hive designs is the Langstroth hive. Named after the inventor, Reverend Lorenzo Lorraine Langstroth, this design revolutionized modern beekeeping. The Langstroth hive consists of rectangular boxes stacked vertically, with removable frames inside. This design allows for easy inspection and management of the hive, as each frame can be lifted out individually. The frames contain a wax foundation, providing a guide for the bees to build their comb and store honey and pollen. The Langstroth hive also includes a bottom board, multiple supers, an inner cover, and an outer cover.

Another popular hive design is the top-bar hive, known for it's simplicity and natural approach. In this design, bars are placed horizontally across the top of the hive, and the bees buld their combs downwards. The top-bar hive mimics the bees' natural behavior, allowing them to create their own comb without the use of a perforated foundation. Additionally, the top-bar hive is often chosen for its ease of construction and maintenance. It requires fewer materials and is less expensive than the other designs. However, it is important to note that managing a top-bar hive can be challenging, as it requires a deep understanding of bee behavior and comb management.

When considering hive design and structure, it is essential to prioritize the well-being and comfort of our bees. A properly designed hive ensures that the bees have sufficient space for brood rearing, honey storage, and pollen collection. It also provides adequate ventilation and insulation to maintain the hive's temperature and humidity levels. Furthermore, the hive should be positioned strategically to optimize sunlight exposure and protection from harsh weather conditions. Placing the hive in a sheltered area can help reduce the impact of wind, rain, and extreme temperatures on the colony.

In conclusion, understanding hive design and structure is paramount for successful beekeeping. Whether you opt for the Langstroth hive or the top-bar hive, ensuring that your bees have a well-designed and carefully constructed home is essential for their health and productivity. By prioritizing their needs and mimicking their natural behavior, you can create an environment where they thrive and flourish. So, delve into the world of hive design and structure, and embark on a journey to become a knowledgeable and skilled beekeeper.

Chapter Fourteen

Beekeeping Equipment

In the fascinating world of beekeeping, having the right equipment is essential to ensure the well-being of your bees and to make your beekeeping journey a successful one. This chapter will guide you through the different types of beekeeping equipment you will need and how to properly use them. As a beekeeper, it is crucial to have a good understanding of the different supplies and tools that are necessary for the successful management of your bee colonies.

First and foremost, every beekeeper needs a hive. A hive serves as a home for your bees, providing them with shelter and a safe place to build their honeycombs. There are various types of hives available, but the most commonly used ones are the Langstroth hive and the top-bar hive.

The Langstroth hive consists of multiple boxes called supers, stacked on top of each other, with removable frames, allowing for easy inspection and maintenance. Within the beehive, frames are used to hold the beeswax foundation, which acts as a guide for the bees to build their honeycomb. These frames can be made of wood or plastic and are easily removable for inspection or honey extraction. The top bar hive features bars placed horizontally for the bees to build their comb on.

Another essential piece of equipment is the smoker. The smoker is used to calm the bees during hive inspections by emitting cool smoke. This smoke disrupts the bees' communication system, making them less likely to become agitated or defensive. The smoker is typically filled with the materials such as pine needles, wood chips, or burlap, which produces a cool, white, smoke. The smoke triggers a response that prompts them to gorge themselves on honey. This engorgement makes them less likely to sting, allowing you to work with them more easily.

To protect yourself from bee stings, it is vital to wear the appropriate protective clothing. A beekeeper's suit usually consists of a full body coverall made of lightweight, breathable material, along with gloves, a veil, and a hat. These items provide a physical barrier between you and the bees, reducing the risk of stings.

A hive tool is another indispensable tool for beekeepers. A hive tool is a versatile instrument. A hive tool with a hooked end is particularly useful for separating frames from boxes. While the flat end cam be used for prying or scraping. It also helps in scraping off excess propolis, a resinous substance that bees use to seal cracks and crevices in the hive.

For inspecting the frames and checking the health of your colony, you need a bee brush. A bee brush has soft bristles that enable you to gently brush away bees from the frames without harming them.

When it comes to harvesting honey, an extractor is an invaluable tool. . This device spins the frames, using centrifugal force to extract the honey from the comb without damaging the comb. Extractors come in different sizes and styles, including manual and electric options, depending on the size of your operation and your personal preference.

Lastly, don't forget to feed your bees. Bee feeders come in various forms, such as entrance feeders, top feeders, and frame feeders. These feeders provide supplemental food, such as sugar syrup or pollen substitute, to your bees during times of scarcity or when you are starting a new colony.

Other tools include a queen excluder and a pollen trap. A queen excluder is used to keep the queen out of certain hive sections. A pollen trap is used to collect pollen from trees.

In conclusion, a beekeeper must have a good understanding of the various supplies and tools used in beekeeping. These include beehives, frames, smokers, protective clothing, hive tools, bee brushes, extractors, excluders, and pollen traps. By investing in quality equipment and using it correctly, you will be well on your way to becoming a skilled and responsible beekeeper

Chapter Fifteen

Bee Disease and Parasites

As we delve into the realm of beekeeping, it is crucial to understand the intricate web of bee diseases and parasites that can affect our beloved pollinators. These tiny creatures, although resilient, are susceptible to numerous ailments that can weaken their colonies and impact their overall health. In this chapter, we will explore the various diseases and parasites that beekeepers may encounter, equipping us with the knowledge and tools necessary to protect our buzzing companions.

First and foremost, let us focus on the diseases that can afflict bees. One of the most common ailments is called American Foulbrood. The bacterial infection is highly contagious and can decimate an entire colony if left unchecked. It is characterized by a foul-smelling brood, which appears dark and sunken. To prevent the spread of American Foulbrood. , it is essential to implement strict hygiene practices, such as regularly cleaning equipment and promptly removing infected brood.

Similarly, European Foulbrood is another bacterial disease that poses a threat to bee colonies. Unlike its American counterpart, European Foulbrood affects the larvae, causing them to get twisted and discolored. Although this disease is less severe, it can still weaken the colony if not addressed promptly. Regular inspections and the removal of infected brood are crucial in managing European Foulbrood.

Another significant disease that beekeepers must be cautious of is Nosema. Nosema is a fungal parasite that affects the digestive system of bees. Infected bees often display symptoms such as diarrhea and a reduced lifespan. To mitigate the impact of Nosema, it is crucial to maintain proper nutrition for the bees and ensure they have access to a clean water source.

Moving on to parasites, the Varroa mite is perhaps one of the most notorious threats to bees worldwide. These tiny, blood-sucking parasites attach themselves to adult bees and their brood, weakening the bees and transmitting viruses. Regular monitoring and treatment for Varroa mites are essential to protect the health and longevity of the colony.

Another common parasite is the Small Hive Beetle. These beetles lay their eggs in the wax combs, and their larvae can cause extensive damage by tunneling through the hive. Vigilance in monitoring and promptly removing any affected combs is crucial in managing this pesky intruder.

Lastly, we must not forget the infamous wax moth. Wax moths lay their eggs in the beeswax, and their larvae consume the comb, leaving behind a sticky mess. Maintaining a clean and well-maintained hive, along with regular inspections, can prevent wax moth infestations.

Understanding the various diseases and parasites that can affect bees is vital for every beekeeper. By implementing proper management techniques and practicing good hygiene, we can minimize the impact of these threats on our bee colonies. Remember, a healthy and thriving hive is the result of our dedication and knowledge in combating bee diseases and parasites.

Chapter Sixteen

Honey Harvest and Extraction

Welcome to the chapter where we delve into the fascinating process of honey harvest and extraction. In this section, we will explore the key steps and techniques involved in harvesting this golden nectar from our buzzing friends, the bees. As beekeepers, one of the most rewarding moments is when we finally get to taste the fruits of our labor by collecting and enjoying the honey that our industrious bees have produced. However, it is crucial to approach the harvest and extraction process with care and respect for both bees and the honey itself.

Before we embark on the journey of honey harvest, it is essential to understand the timing and readiness of the honeycombs. Bees work tirelessly to fill their hexagonal cells with honey, and it is vital to ensure that the honey has ripened adequately and that the comb is sealed with wax caps. This ensures the honey is at its peak in terms of flavor, consistency, and moisture content. When the time is right, it is crucial to prepare ourselves with the necessary equipment. A beekeeper's toolkit for honey harvest typically includes a bee suit, gloves, a bee brush, a smoker, a hive tools, and, most importantly, the honey extractor. The honey extractor is a device designed to remove honey from the comb without damaging it, allowing us to collect the honey while leaving the comb intact for the bees to reuse.

To begin the harvest, we first need to gently smoke the hive to calm the bees. This helps reduce their defensive response and allows us to work peacefully. With great care and precision, we then remove the honey supers, which are the boxes placed on top of the brood chamber specifically for honey storage.

Once we have the supers in our possession, it's time to head to the honey house or designated extraction area. Here, we set up our extraction equipment and begin the process of uncapping the honeycomb cells. Uncapping involves removing the thin layer of wax that seals each cell, revealing the sweet honey within. To uncap the honeycomb, we can use either a hot knife or the uncapping fork. The hot knife gently melts the wax as it glides across the cells, while the uncapping fork allows us to scrape off the wax caps manually. Both methods are effective, and the choice depends on personal preference and available equipment.

Once the cells are uncapped, we place the frames into the honey extractor. The device uses centrifugal force to spin the frames, causing the honey to be flung out of the comb and collected at the bottom of the extractor. This process is gentle on the comb and ensures that the bees can reuse it, ultimately reducing their workload. As the honey collects at the bottom of the extractor, we have the option to filter if desired.

Filtering helps remove any small particles such as wax fragments or bee parts, resulting in a clearer and more visually appealing honey. However, it is important to note that filtering is not necessary and can impact the texture and flavor of the honey. Once the honey has been extracted and filtered, it is time to bottle and store it properly.

Clean and sterilized jars or containers should be used to preserve the honey's quality and prevent any contamination. Remember to label the jars with the date and batch information, allowing you to keep track of your harvest and maintain a fresh stock.

In conclusion, honey harvest and extraction is a crucial aspect of beekeeping with the right tools, knowledge, and respect for the bees, we can enjoy the fruits of their labor while ensuring their well-being. So, let us embark on this delightful journey of harvesting and extracting honey, appreciating the incredible work of our honey-making friend, the bees.

Chapter Seventeen

Swarm Control and Hive Splitting

Swarm control and hive splitting are two essential techniques in beekeeping that every beekeeper should be familiar with. These methods not only help maintain the health and vitality of the hie but also provide opportunities for expansion and growth. In this chapter, we will explore the importance of swarm control and hive splitting , as well as the steps involved in successfully implementing these techniques. Swarm control is the process of managing the natural instinct of honeybees to reproduce and form new colonies. When a hive becomes overcrowded or lacks sufficient resources, the bees may decide to swarm., which involves a large group of bees leaving the hive in search of a new home. While swarming is a natural behavior, it can be detrimental ot a beekeeper, as it results in the loss of honey production and potentially weakens to original hive. Therefore, it is crucial to implement swarm control measures to prevent or manage swarming

One effective method of swarm control is known as hive splitting. This technique involves dividing a healthy and thriving colony into two or more separate hives. By doing so, the beekeeper creates additional space for the bees, reducing the likelihood of swarming. Hive splitting also serves as a means of increasing the number of colonies, and allowing for the potential expansion or creation of new hives for other purposes, such as queen rearing to successfully control swarming and implement hive-splitting, certain factors must be considered.

First and foremost, timing is crucial. It is essential to monitor the hive closely and identify any signs of swarming preparations, such as the presence of queen cells or an increase in bee population. By recognizing these cues, the beekeeper can intervene before the swarm departs.

Once the decision to perform hive splitting has been made, the beekeeper needs to prepare the necessary equipment. This includes additional hive bodies, frames, bottom boards, covers, and a means of transporting the newly separated colonies. It is essential to ensure that the equipment is clean and in good condition to provide a suitable environment for the bees. The actual process of hive-splitting involves carefully dividing the frames and bees between the new hives. The queen is typically placed in one of the hives, while the other hive receives a new queen or queen cells. It is crucial to handle the frames and bees with care to minimize stress and potential damage. Additionally, proper feeding and monitoring of the newly split hives are necessary to support their establishment and growth.

Swarm control and hive splitting are not only vital techniques in beekeeping but also offer valuable opportunities for beekeepers to manage and expand their colonies. By understanding the natural behavior of honeybees and implementing these methods, beekeepers can maintain healthy hives, prevent swarming, and potentially increase their bee populations.

Don't miss out!

Visit the website below and you can sign up to receive emails whenever Linda Standridge publishes a new book. There's no charge and no obligation.

https://books2read.com/r/B-A-IMCBB-ATSPC

BOOKS 2 READ

Connecting independent readers to independent writers.

About the Author

Linda Standridge is a small-town southern gal. She spends her days fantasizing about sexy aliens or simply writing about sexy aliens. When she isn't doing that she is spending time with her grandchildren and her children. She enjoys gardening and crocheting too. She promises she hasn't lost it entirely. She just loves talking about herself in the third person. She'd love to hear from ya'll.

Give her a shout-out at lindastandridgeauthor@gmail.com